What Is Red?

by Jade Bibocul

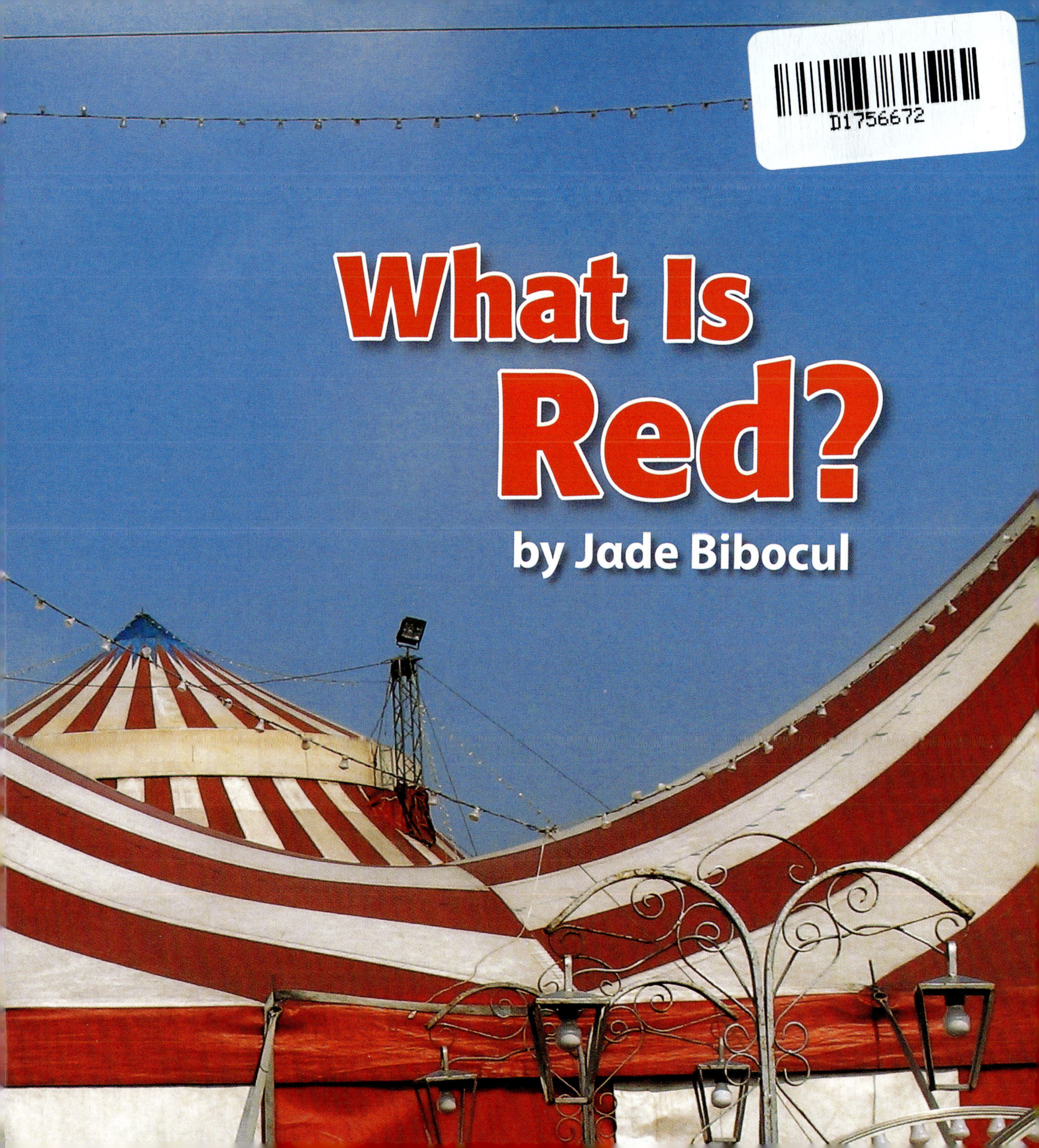

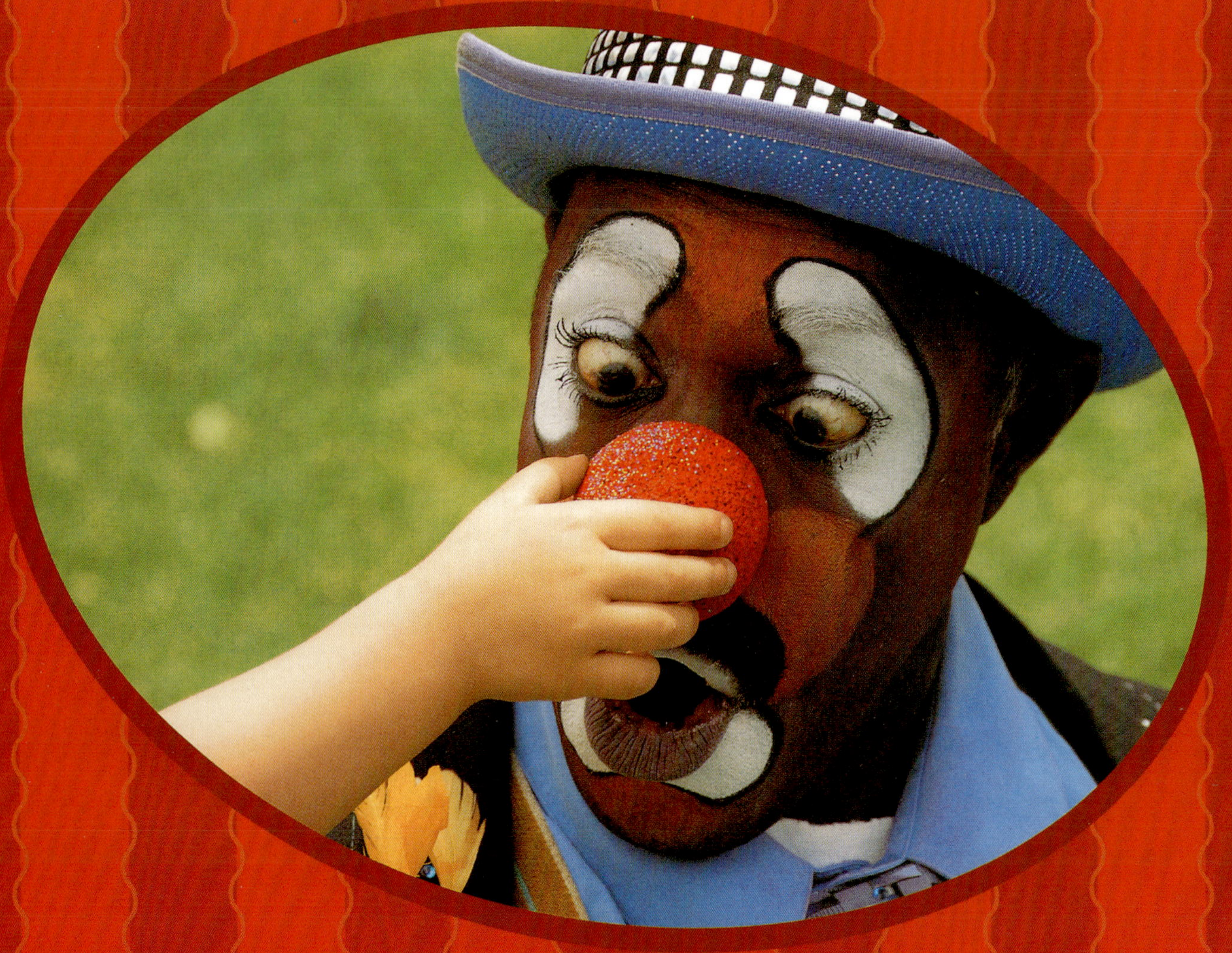

round red nose

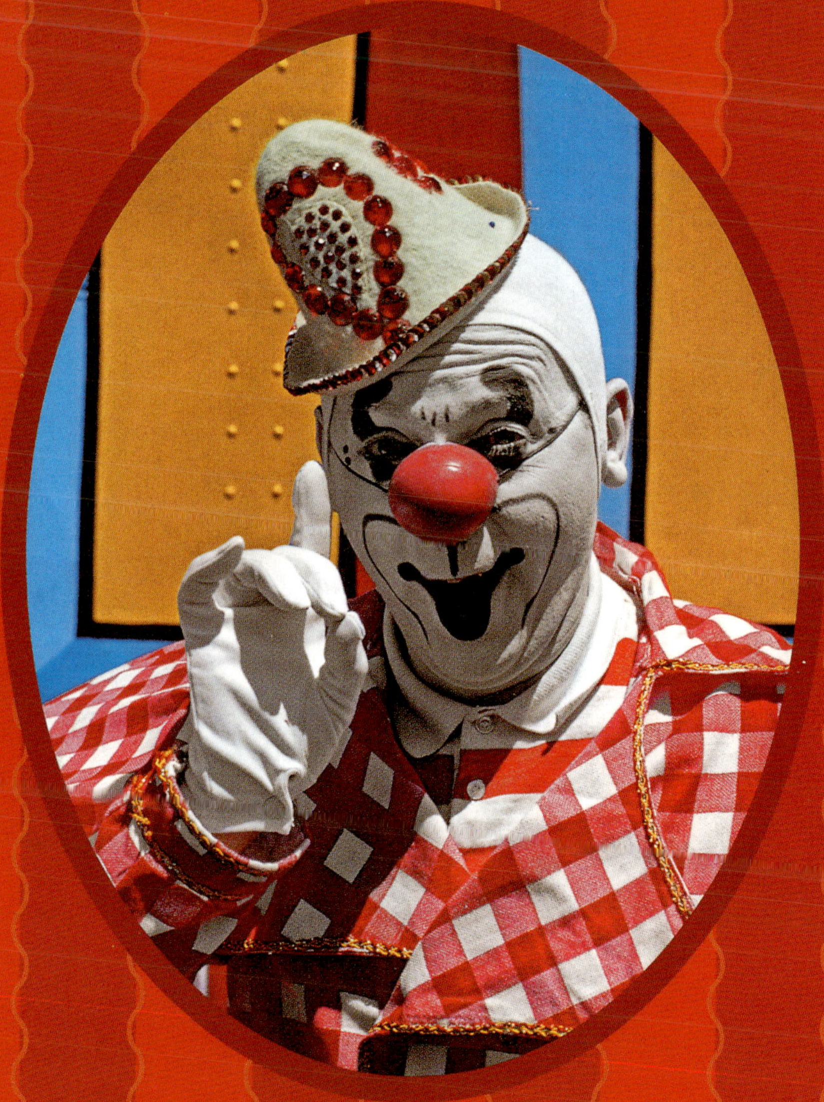

square red pattern

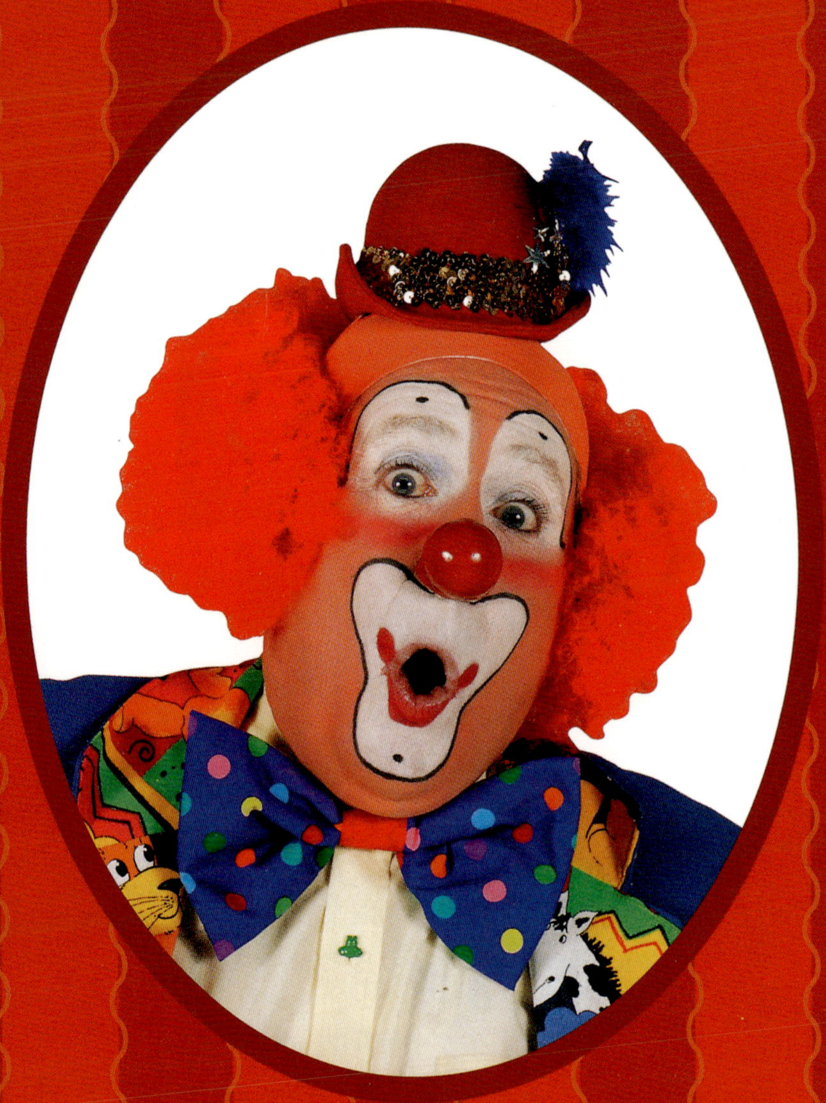

small red hat

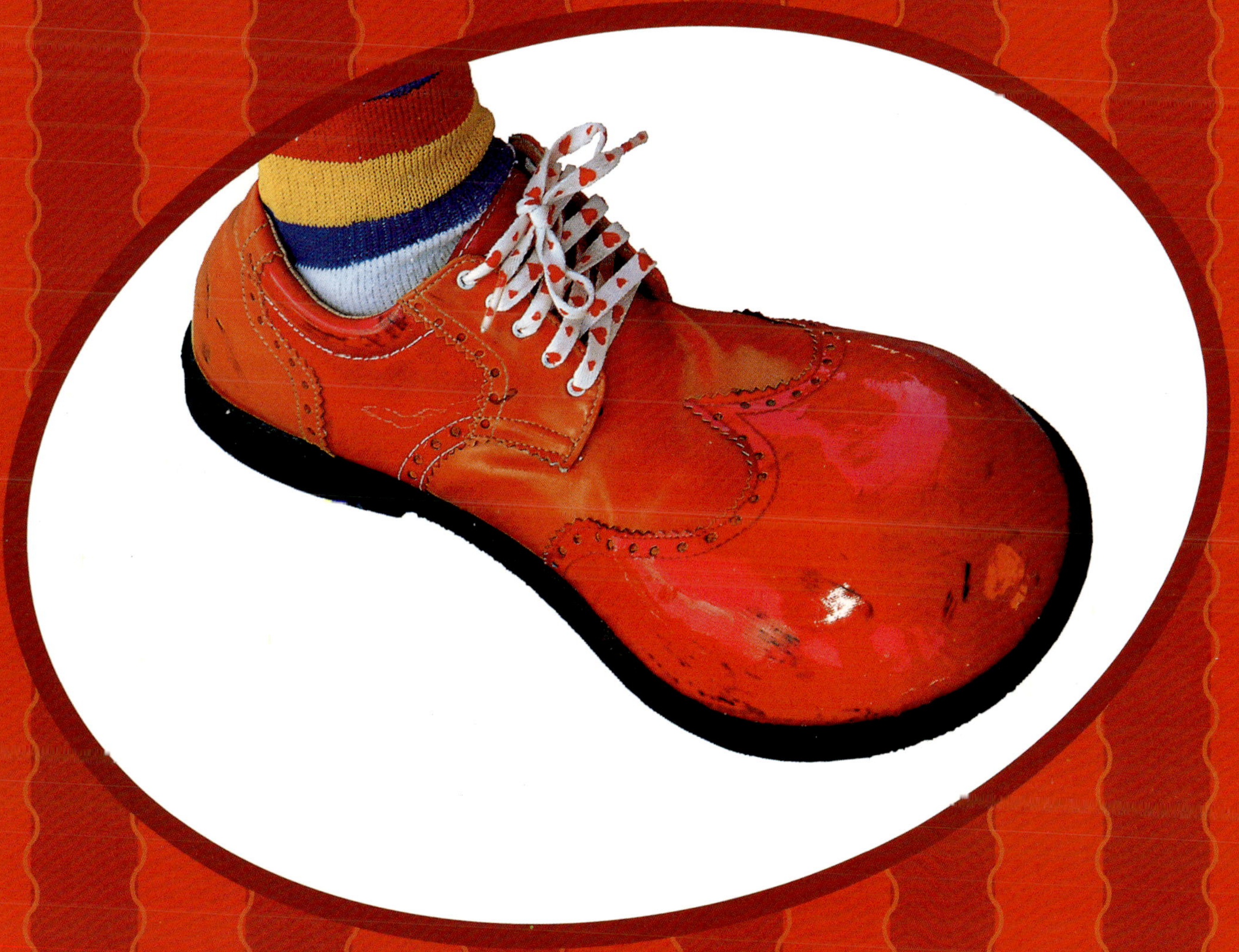

big red shoe

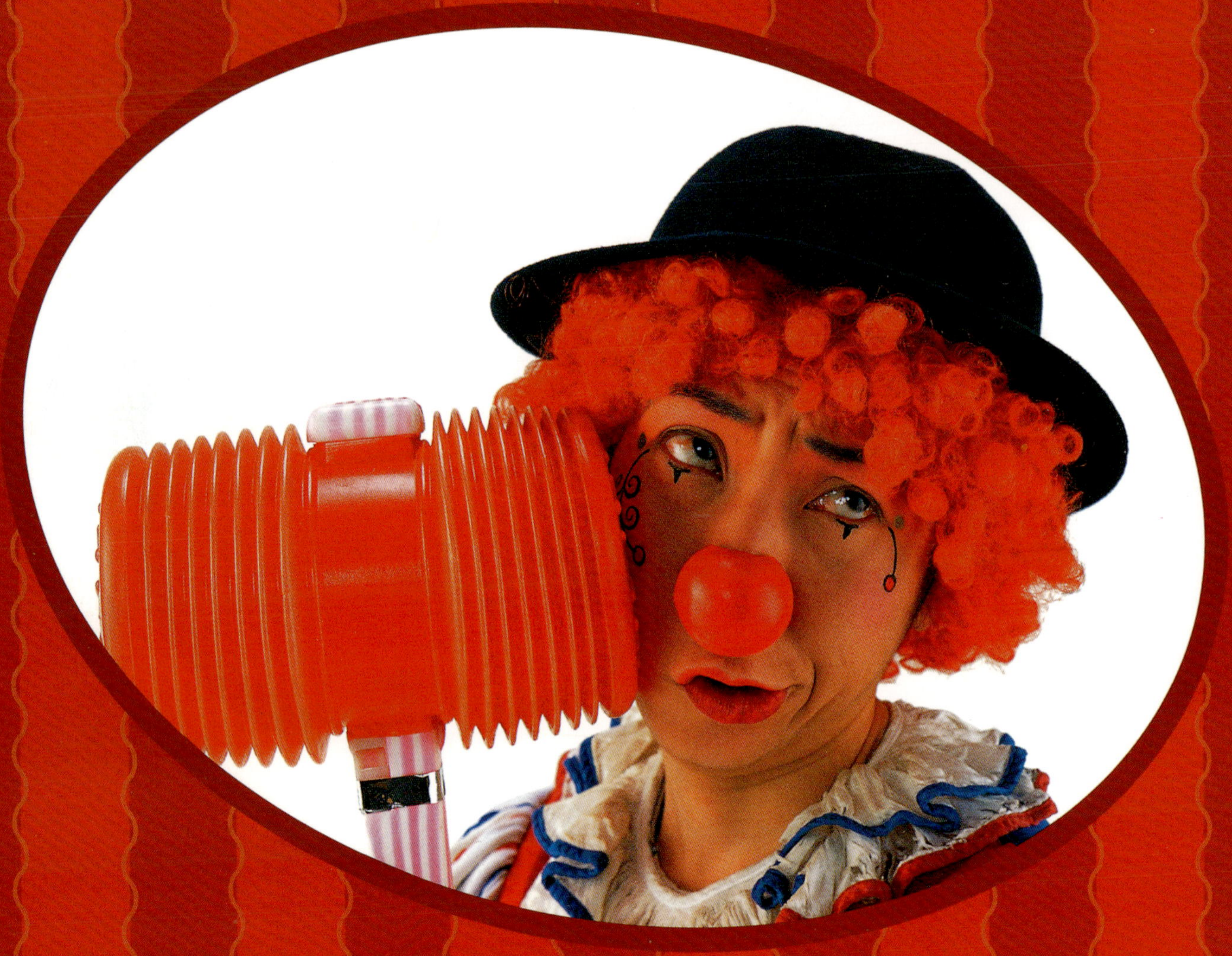

rough red hammer

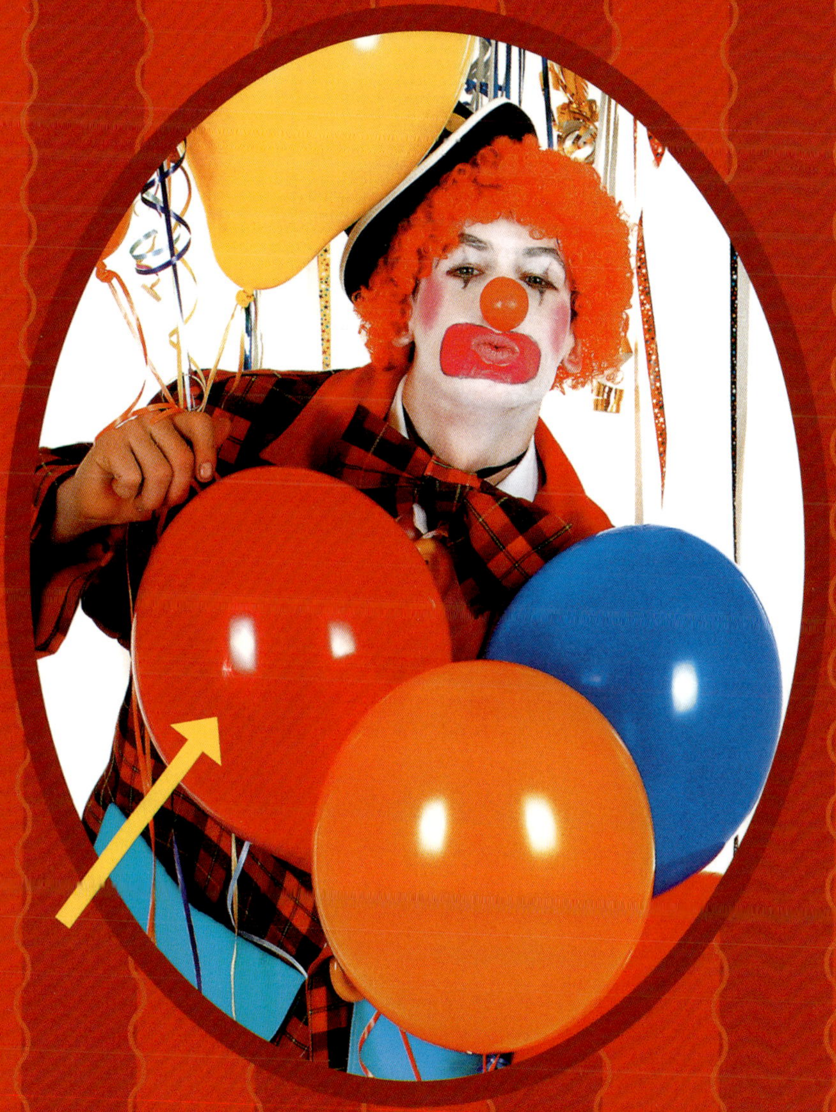

smooth red balloon

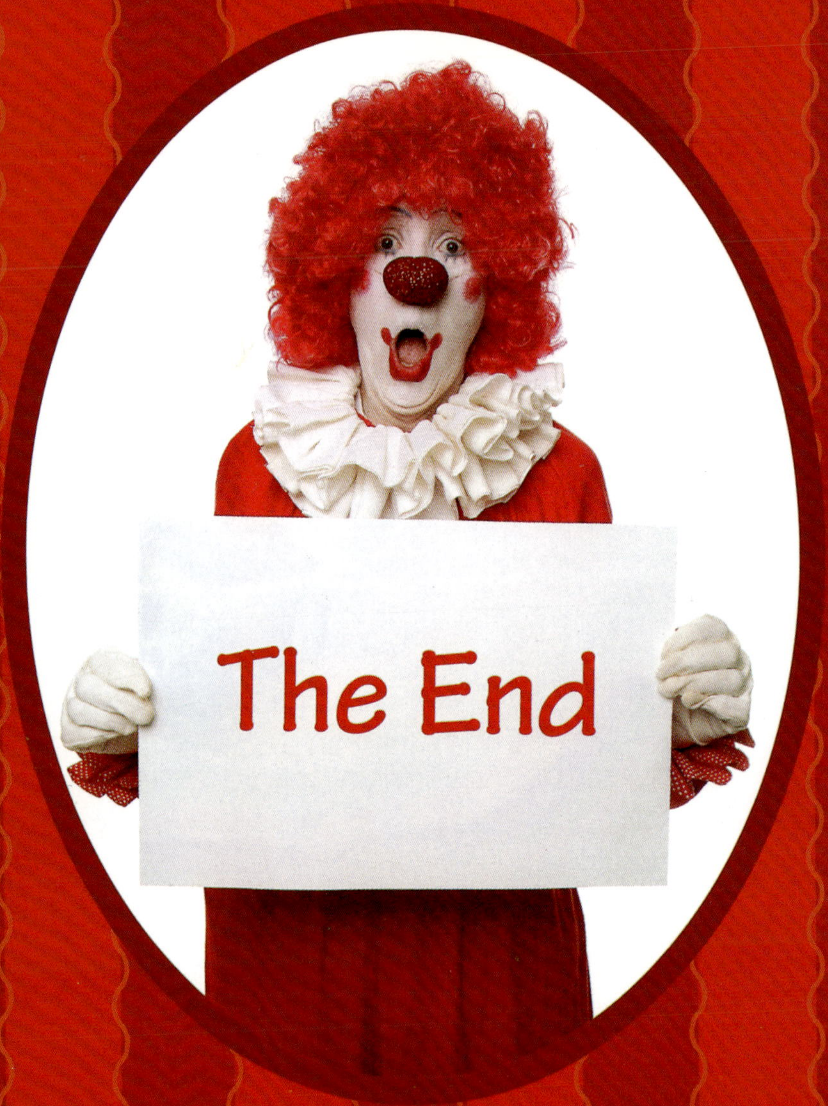

two red words